ANIMALS EXPOSED!

The Truth About
Animal Builders

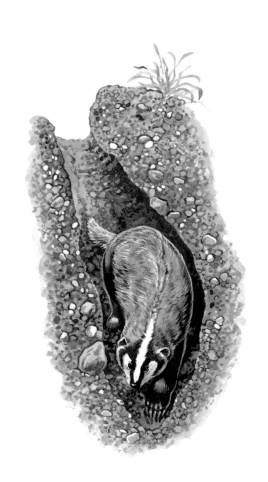

Created and produced by Firecrest Books Ltd
in association with the
John Francis Studio/Bernard Thornton Artists

Published by Tangerine Press,
an imprint of Scholastic Inc;
557 Broadway, New York, NY 10012

ISBN 0-439-54328-2

Printed and bound in Thailand
First Printing June 2003

ANIMALS EXPOSED!

The Truth About Animal Builders

Bernard Stonehouse
and Esther Bertram

Illustrated by
John Francis

Tangerine Press and associated logo
and design are trademarks of Scholastic Inc.

FOR ADAM

Art and Editorial Direction by
Peter Sackett

Edited by
Norman Barrett

Designed by
Paul Richards, Designers & Partners

Color separation by
**SC (Sang Choy) International Pte Ltd
Singapore**

Printed and bound by
Sirivatana Interprint Public Co., Thailand

Contents

Introduction   6

Diggers   8

Burrowers   10

Weavers   12

Plasterers   14

Builders   16

Jet Engineers   18

Gliders and Fliers   20

Spinners   22

Armor Plating   24

Camouflage   26

Stitchers and Stickers   28

Packagers   30

Defense Masters   32

Sailors and Diver   34

Chompers and Grinders   36

Drillers   38

Carpenters and Woodworkers   40

Built-in Tool Users   42

Tool Users and Makers   44

Temperature Controllers   46

Index   48

Introduction

Wherever we live, we're surrounded by walls, houses, furniture, roads, and machinery, which builders and engineers create to make our lives safer or easier. Building and engineering are not only for humans, animals are engineers and builders, too. Thousands of different kinds of animals are into digging, mining, building, weaving, and carpentry – making simple or sometimes very complicated structures for their comfort and protection, just as we do.

From early childhood we grab sticks, stones, boxes, and toy bricks to make them into other things, mainly for fun and to learn how to do it. Animals do the same, but right from the start it's for real. Diggers dig holes to live in or store food in, and tunnels for running to safety. Miners mine the soil for earthworms, grubs, beetles, and grass roots to feed on. Nest-builders build homes for their eggs and young, using sticks, grass, and mud. Different kinds of animals create nets, blankets, houses, cradles, even rock walls hundreds of miles long, using the teeth, claws, bricks, threads, and other tools, that nature provides.

There are differences. Human builders plan the kind of structures they want, then employ skills and experience to build them. Very few animals plan ahead like this. Their building comes mainly from instinct, though it is sometimes improved by practical experience – for example, a spider's second or third web is better than its first. This book shows what animal builders and engineers can achieve, using their natural skills and raw materials.

Beavers are great examples of animal builders. They gnaw down young trees with their sharp teeth, eat the sweet bark, and use the trunks and branches to build their homes, or lodges, and to dam streams.

Diggers

Left: A chipmunk digs a hole to bury its spare nuts.

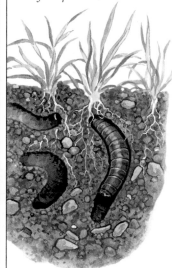

When we take a spade and dig a hole, it is usually to bury something, or to dig up something else that might be there. This chipmunk (opposite) is digging a hole with its forepaws. In a moment, it will roll those nuts into it and bury them. Then, in winter, when no nuts are left on the trees, it will come back, dig the nuts up, and eat them.

So a clever chipmunk plans for the future? Not really – a chipmunk is not that thoughtful. In autumn, when the trees are full of nuts and the chipmunk has eaten all it can, it digs shallow scrapes and buries spare nuts. In winter, when it's hungry, it will scratch around its territory for food, and may dig up some of the nuts and eat them. The chipmunk doesn't have a checklist, and will forget where many of them are. But burying a few nuts when they are plentiful can help to keep it going through a hard winter.

Underground living

Larvae of many kinds of moths and beetles live in soil. Adults lay eggs just below the soil surface. The larvae develop, digging their way through the soil in search of smaller animals, grass roots, and other plants on which they feed. Some spend two or three winters below ground before emerging as adults.

Digging tunnels

An earthworm lives in soil and digs long tunnels with its nose. Without hands or feet, the earthworm presses the tip of its nose between the soil grains, then pumps blood into the head area to force them apart, making room for the rest of its body. In this simple way, earthworms dig miles of tunnels through the soil in search of food.

Digging for food

Gulls and rooks dig with their bills for earthworms, insects, and other small soil animals. Digging in hard-packed earth is tough. But it is much easier if the soil surface is broken by plowing. Watch out in winter for flocks of birds, like these black-headed gulls, that follow the plow and dig for food in the freshly turned earth.

Digging a pit

The ant lion's method for catching ants is to dig a pit and wait for them to fall in. Ant lions, the larvae (young forms) of flying insects, dig steep-sided conical pits in sandy soil, partly burying themselves at the bottom. When an ant or other small insect rolls down the slope, the ant lion is waiting with jaws open.

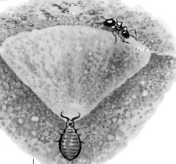

Burrowers

Opposite page: This female polar bear, sleeping in a snowdrift burrow, has made a warm den for herself and her tiny cubs.

If you live in the Arctic in winter, with an air temperature of −20°F (−29°C) and nothing but ice and snow for miles around, where is the best place to keep warm? Try burrowing in a snowdrift, with a sleepy polar bear for company.

Snow is a good insulator. Inuit, or Eskimo, hunters build shelter houses of ice and snow. The polar bear, with a body temperature close to our own, keeps the inside of its den warm and comfortable. That is how polar bear cubs survive. In early winter, the mother bear burrows into a hard-packed snowdrift, making a cave in which she curls up and sleeps. More snow settles over the drift. In early spring, she awakens long enough to give birth to two or three rabbit-sized cubs, which find her milk glands and start to feed. For the next 10 to 12 weeks the cubs feed sleepily, kept warm by their mother's huge body. In late spring, they all waken, and the mother leads her family out into the cold Arctic world.

Burrowing in a termite mound

Snakes can burrow. This Indian cobra has burrowed into a termite mound, pushing aside the eggs and larvae of thousands of antlike termites, and laid its own clutch of a dozen white-shelled eggs. The termites cannot harm the eggs, and will help to defend them with their own, keeping them warm and safe until they hatch.

Burrowing badgers

Badgers live in burrows which they dig using their strong front paws. Starting at a bank or tree root, they burrow furiously with their forelimbs, pushing the earth back behind them. Often they clear and extend old holes to make underground warrens. A burrowing badger can shift more than a ton of earth a day.

Cliff burrows

Sand martins burrow in cliffs. These have found a cliff of soft, sandy shale, and burrowed with bills and feet to scrape away the loose material. Each pair will dig a burrow a foot or more long, just wide enough for them to slip through, with a nest chamber at the end. That will make a safe family home, usable for several years.

Burrowing bees

Can a bumblebee burrow? Yes, quite easily in soft ground, to make a home for its eggs and young. The queen bee burrows and removes the soil with her jaws, making a cave 2-3 inches (5-8 cm) deep. A ready-made mouse hole will work just as well. In the burrow, she lays eggs and raises her family.

Weavers

Opposite page: These tiny hummingbirds, only 2 to 3 inches (5-8 cm) long, build a delicate nest of moss and lichen fragments, bound together with strands of spider's web.

Simple nests

How does a bird build its nest? Usually it crouches on a branch or tree fork, settling and weaving the nest material around it, shaping the bowl with its

feet. Some birds, like this hen chaffinch, weave in lichen and moss, and line the nest with feathers. Building will take her four to ten days.

Mammal weavers

Small mammals can be weavers, too. This female harvest mouse has woven a 3-inch (8 cm) diameter nest of split grass and wheat blades. Close to the ground, and anchored among wheat stalks that keep other mammals away, the nest will provide a warm, secure home for successive litters throughout the summer.

Can you weave cobwebs? Maybe not, but a hummingbird can. This tiny white-eared hummingbird of South America has built a miniature nest of moss fragments and lichens, plucked from the trees and held together by a network of cobwebs – strands from a spider's web. Less than 1 inch (2 cm) across, the nest holds two eggs smaller than your fingernail. In three weeks, these will hatch into tiny, naked chicks. Reared mainly on nectar, the chicks will be ready to fly just three to four weeks later. Several other kinds of hummingbirds weave cobwebs into their nests. Though fine and delicate, the web strands are slightly sticky and surprisingly strong – just right for holding a small nest together.

Why are they called hummingbirds? Because their wings beat fast like those of bees or mosquitoes, making a quiet but persistent hum. The fast wing beat allows them to fly backward as well as forward, and even hover in midair. They feed by hovering to take nectar from flowers.

Hanging nests

Birds nesting in tropical forests need to protect their eggs from predatory snakes and lizards. Village weaverbirds of Africa build hanging basket nests on the outermost tips of branches. Cocks start each nest as a globe of interwoven loops, then attract a hen that takes over, completes the nest, and raises a family.

Basket weaving

Red-headed malimbes of southern Africa use stiff grasses to weave the most elegant of hanging nests. Each nest is a covered basket with a narrow tunnel entrance. The hen incubates comfortably in her basket, shaded from the sun and rain. The hanging tunnel entrance makes it almost impossible for snakes and other predators to find a way in.

Plasterers

Narrowing the doorway

Nuthatches of Europe and North America breed in hollow tree trunks and cavities. When nesting birds find a cavity, perhaps an abandoned

woodpecker's nest, they plaster up the entrance hole with mud, making it too narrow for woodpeckers or any other large birds to enter. That way they secure the cavity for themselves.

Did you ever make a clay pot or mud pie? Then you'll know that mud and clay can be molded into shapes that harden when they dry. Birds and many other animals know this too, and some find it useful in nest-building.

Small birds weave small nests that hang together comfortably. Bigger birds need bigger nests, made from heavier materials and held together with something stronger than cobwebs (see pages 12–13). These two cormorants have started to build a nest on a high, windy cliff above the sea. The hen sits defending the nest site from other cormorants. The cock flies in with beakfuls of seaweed and mud from the shore. He passes it to the hen, who arranges it round her. Without the mud, the seaweed would blow away. Without the seaweed, the mud would wash off. Together they harden into a strong, bowl-shaped nest, sturdy enough to hold four or five eggs. Later, with patching and mending, it will be home to a family of lively chicks.

Under the eaves

In the wild, swallows nest on cliffs. In towns and villages, they nest under eaves – the overhanging edges of roofs, which protect them. The nests are cups made of mud, which they gather in their beaks from stream banks, and build out a little bit at a time from the wall. A well-built nest will still be there next spring.

Opposite page: These common cormorants are building a nest of seaweed and mud, in which the hen will lay her eggs.

Mud pies

Flamingos nest in colonies of thousands around lake edges. Their nests are made of mud, scraped and gathered from the lake floor and plastered into piles 12 to 15 inches (30-40 cm) high. Hot sunshine bakes the mud hard enough to withstand rainfall and repeated flooding. The parents incubate in turns, perching awkwardly with long legs folded until the single chick hatches.

Walling in

When a hen hornbill starts to lay her eggs in a hollow tree nest, the cock walls her into the cavity. He plasters over the whole of the entrance except for a narrow slit, through which he feeds the hen through six or more weeks of incubation. When the chicks hatch, he feeds them too, until all are old enough to come out.

Builders

Coral reefs

Coral reefs are undersea walls in tropical seas, some of them hundreds of miles long, built from the bodies of millions of tiny animals. Each animal, like a minute sea anemone, builds within itself a hard, chalky skeleton. Over thousands of years the coral animal skeletons build up to form the huge, solid walls we call coral reefs.

Tower builders

Termites are antlike insects that live in large colonies in warm and tropical countries. Different kinds of termites build different kinds of nests, some underground, others on the surface. These Australian desert termites mix their own droppings with sandy soil to create tower nests 10 feet (3 m) high, enclosing galleries and chambers for thousands of worker and soldier termites.

Making an open clay pot (see pages 14–15) with your fingers is one thing, but how would you build a completely round, hollow clay pot the size of a basketball, with only a beak as a tool? That is what South American ovenbirds do. The birds pair up and start building in winter. This is long before they need the nest, but winter is the rainy season, when there is plenty of soft mud around. They start with a cup nest, then build up the walls, one bird bringing the mud, the other plastering and shaping from inside. When the walls reach head-height, they complete the domed roof, and some pairs add a curved tunnel to the entrance. The mud hardens to a one inch-thick shell, strong enough to protect the family from rain, wind, and the searing heat of summer.

They are called ovenbirds because the completed nest, perched in trees, on telegraph poles, and wall-tops, remind people of old clay ovens.

Right: Round or oval mud nests of South American ovenbirds are built in late winter, in plenty of time for the spring breeding season.

Making a swimming pool

Frogs need to protect their eggs and tadpoles from drying out. This blacksmith frog, living on a muddy creek, molds a circular mud bank high enough to hold a pool of water, and lays its eggs in it. Fish cannot enter to eat the eggs. If the creek dries up, the eggs and tadpoles survive in their own private swimming pool.

Dam builders

Beavers cut down young trees to feed on the soft sapwood. In stream valleys, the fallen trees form dams, which hold back the water and create lakes. This is where beavers build their lodges – houses made of branches and mud, where they raise their young. If the dam leaks, the beavers instinctively pack in more sticks and mud to stop it.

Jet Engineers

Left: This archerfish is aiming a jet of water at an insect on the reeds. The insect will be washed off and eaten.

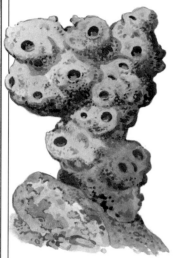

Hydraulics is a branch of engineering that harnesses the power of moving water and other liquids, involving jets and currents. Plants use hydraulics to keep their shape, and to carry food between leaves, stems, and roots. Animals use streams of water internally (for example, in the blood systems). Many that live in water use streams and jets to move themselves around.

Are animals hydraulic engineers? Sure, if we want to give them a long name. Here is a fishy engineer, using hydraulics to catch its supper. The archerfish, about 10 inches (25 cm) long, lives in streams and mangrove forests along the coast of Indonesia. Regular archers shoot arrows. Archerfish blow a stream of water droplets at insects on or above the water surface. Forward-looking eyes and a jet-shaped mouth help them take aim. Pressure from the gill covers produces a jet up to 12 feet (nearly 4 m) long. An insect bowled over by the stream is eaten before it has time to recover.

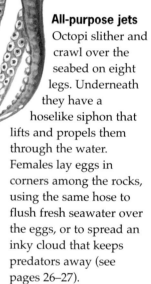

All-purpose jets

Octopi slither and crawl over the seabed on eight legs. Underneath they have a hoselike siphon that lifts and propels them through the water. Females lay eggs in corners among the rocks, using the same hose to flush fresh seawater over the eggs, or to spread an inky cloud that keeps predators away (see pages 26–27).

Channeling food

The sponges you use in your bath were once living animals. It may be hard to believe, but natural sponges are skeletons of animals that once lived on the sea floor. Inside live sponges are channels lined with tiny cells, some moving water currents through the body, others grabbing tiny food particles as they drift past. Collected by a diver, this one was dried in the sun and cleaned to make a bath sponge.

Jet propulsion

Scallops (below left) are slim, double-shelled mollusks related to mussels and oysters. They live on the seabed. Some kinds attach themselves to rocks, but most scallops swim actively in search of food and to avoid predators. The edges of the shells open, then snap together quickly, producing jets of water that drive the scallop forward, backward, or sideways through the sea.

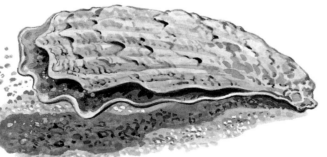

Pumping water

Oysters (above right) are built like scallops, but have a sedentary way of life. They attach themselves to rocks, and live by pumping water into and out of their body. The circulating water brings oxygen to the gills and scraps of food that the oyster traps. If the tide goes out or other dangers threaten, the shells clamp shut until conditions are safe again.

Gliders and Fliers

Birds fly; and so do many insects and a few kinds of mammals. Would you believe in flying fish? Flying frogs? Flying squirrels? They have all been talked about, but do they exist?

Years ago, country folk in Asia and South America told animal collectors about frogs that flew from trees. It took many years to sort out the truth. We know that frogs hop and have webbed feet for swimming, and that some even climb

trees. But can they fly? Well, here is a small, lightweight tree frog, with unusually long webbed fingers and toes. It climbs trees, hopping and crawling along the branches, and feeding on insects as it goes. Then it hops off of a branch, spreading hands and feet to make four "parachutes," one at each corner, that steer and control its descent. It's fun and a good way of getting to the bottom of the next tree, but in a bird we would call this gliding, so it's a gliding frog.

Insect migrants

North American monarch butterflies are famous for their long-distance migration from Canada and the Northern United States to Mexico. Females lay eggs as they go, and in the Spring the new generation joins the older ones in moving back north. Unlike many insects, which are simply transported by the wind, Monarchs seem to be able to fly against light winds as they make these huge journeys.

Flying squirrels

Sitting upright on a branch, this little animal looks very much like any other squirrel. When it leaps, the four limbs spread wide, with furry skin extending between them to form a kite-shaped wing. Flying squirrels can soar and change direction in flight, and make safe, controlled landings on tree trunks – another example of gliding, but not flying.

Opposite page: With long, webbed fingers and toes, this tree frog can turn both hands and feet into parachutes to brake its descent.

True fliers

Albatrosses show the rest of the world how to fly. Biggest of flying birds, with wings spanning 11 feet (over 3 m) and permanently outstretched in flight, wandering albatrosses are able to sustain an effortless gliding flight over hundreds of miles. Their wings are most effective in strong winds. When there is no wind, they struggle to take off.

Flying fish

These tropical fish feed in shoals at the sea surface, leaping into the air when chased by tuna and other predators. With the long tail fin pushing and the long pectoral fins taking their weight, they rise 10-20 feet (3-6 m) above the waves, leaving pursuers far behind. Flying fish can steer 100 feet (30 m) or more across the wind before dropping back into the sea. But, again, it's gliding, not flying.

Spinners

Spiders produce fine strands of strong, silky material from glands on their bodies, and they use them in all kinds of ways. This garden spider is using its silk to build a web. Is it very clever to be able to build such a neat one? Not really – clever would mean thinking things out, and spiders don't think like that. They follow instinct that give what looks like very impressive results.

How does a garden spider build its web? It starts with an outer square or triangle of strong threads, from which it runs a cartwheel pattern of temporary threads to a central platform. Starting from the framework, and using the spokes and the platform to walk on, it lays down the spiral of sticky threads that are the main snare. Then it makes a walkway from the central platform to a hiding place, where it lies in wait. When a trapped insect shakes the web, the spider feels the vibrations and runs along the walkway to investigate. Binding the insect with more sticky threads, it injects poison to kill it.

Spinning anchors
Every mussel in a tightly packed mussel bed is held closely to the underlying rock by a bundle of strong black threads. A tiny young mussel, settling for the first time, anchors itself by a single strong thread that keeps it from being swept away by currents or waves. As it grows, it spins more threads to hold it in place.

Air transport
Several kinds of spiders spin lightweight threads that support them in drifting from one place to another. These newly hatched spiders line up along a twig, each spinning a fine thread that floats out behind it. The wind catches the threads, lifts the young spiders off, and carries them to new homes and hunting grounds.

Spinning and knitting
The larvae of caddis flies spin fine silky strands from glands on their body. Some use them to make protective cases, building in sand grains, bits of vegetation, and other materials. Others, like these net-maker caddis larvae, knit the strands into tiny fishing nets, which they use to trap food from the surrounding water.

Spinning silk
We call them silkworms, but they are larvae or caterpillars of moths. Now living only in captivity, the caterpillars are fed on the leaves of mulberry trees. When mature, each spins an outer coat of threads from head glands. Silkworm farmers kill the cocoons and unwind the threads, which make silk for spinning, weaving, and knitting into cloth.

Armor Plating

In the Middle Ages, specially skilled blacksmiths called armorers designed and created jointed metal armor for knights to wear in battle. A great idea? Yes, but not a new one. Lobsters, crabs, and crayfish have worn armor for many millions of years.

This lobster's armor is a covering of chalky plates, toughened with a horny material called "chitin." A single solid plate covers the main body.

The tail is made up of smaller plates, hinged to each other by leathery joints, and the legs and antennae are encased in jointed cylinders. The joints allow movement, but keep the seawater out. The armor is both a protective covering and an outside skeleton (exoskeleton), to which the lobster's muscles are attached inside.

Lobsters grow bigger each year. Every few months, the lobster's shell will split and shed. It will grow a new one underneath to replace the old one. You should watch out for a lobster's claws. They are also armor plated and can pinch.

Growing shell

Conches are marine snails of tropical seas. This queen conch lives in shallow water off Florida and the Caribbean islands. It emerges partly from the opening of its hard, protective shell to feed on seaweeds. Each year it adds a little more to the shell, which can grow up to 1 foot (30 cm) long.

Pointed scales

Fish are covered with protective scales. Those of bony fish are usually flat and horny, overlapping like roof tiles.

Sharks and rays carry thousands of these bony, sharply pointed scales, set in a tough, leathery skin. This makes a flexible armor plating that few other fish or marine mammals can bite through.

Opposite page: Lobsters and many other crustaceans are covered with jointed plates that act as a skeleton and help to protect them from predators.

Suit of scales

Snakes must be able to twist, turn, and strike at their prey – like this Brazilian boa, lying in wait along a tree branch. Their armor is a lightweight suit of tiny scales, set edge-to-edge in tough skin. As the snake grows bigger, it sheds its old skin (almost in one piece) to expose fresh, shiny new skin.

Bony plates

Mammals are usually furry. But this nine-banded armadillo looks more like a lobster. Its surface is made of bony plates set in a covering of rough, horny skin. The nine bands along the back and abdomen make it flexible enough to run and burrow. Very few predators can pierce this armadillo's tough armor plating.

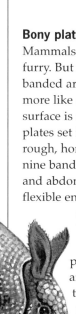

Camouflage

Now you see it, now you don't – camouflage is the craft of disguising or concealing things to match their backgrounds. Soldiers wear camouflaged uniforms, live in camouflaged tents and buildings, and drive camouflaged vehicles. Some animals use camouflage in similar ways.

Colored very much like their background, sitting or standing perfectly still, they are almost impossible to see. Birds, especially, are masters of camouflage. You need to look twice at this picture to see the wryneck resting on the tree trunk. It perfectly matches the bark in color and texture. More important, while we are near, it will stay very still, watching us closely. Then it will slip quietly around the tree trunk to safety.

So wrynecks know all about camouflage? We cannot say what any bird knows. We can only say that, apart from the closely matching feathers, wrynecks have a built-in behavior pattern – standing still when alarmed – that makes their camouflage doubly effective.

Why wryneck? "Wry" means "twisted." The wryneck can twist its neck almost 180 degrees.

Hiding the home
Ready to nest, this European mason bee has found an abandoned snail shell. Packing the peak of the shell with a store of nectar and pollen, it lays tiny, oval eggs, then walls off the entrance with a mixture of sand grains, chewed leaves, and saliva. Finally, it camouflages the shell under a cover of straw and other debris.

Hiding in seaweed
Coastal sea floors are littered with scraps of seaweed and other debris. This sea urchin, scavenging for food, seems to have covered itself with fragments of weed that will help hide it from enemies. Did the urchin gather them, or just let them settle? We don't know, but they often wear camouflage like this.

Warmth and cover
A mother eider occasionally needs to leave her eggs to nibble the grass. On the cold Arctic tundra, the eggs chill if left for long, or a predator might see them. Eiders and some other ducks pull a corner of the downy nest-lining over the eggs, a simple act that both keeps them warm and hides them.

Smoke screen
A small octopus sits among the rocks. It has changed color to match its background, but an inquisitive fish has smelt it, and is about to pounce. The octopus has one trick left – a camouflage screen. From an internal store, it releases an inky black liquid, forming a cloud that confuses the fish, and allows the octopus to get away.

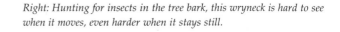

Right: Hunting for insects in the tree bark, this wryneck is hard to see when it moves, even harder when it stays still.

Stitchers and Stickers

Left: This long-tailed tailorbird from southeast Asia has stitched leaves together with cobwebs and grass stems, making a secure bag nest for his mate.

Drab little birds with monotonous voices, the tailorbirds of India, southeast Asia, the Philippines, and Australia are remarkable for one thing – their stitching. Does the hen sit and sew? No, it is more like stapling than sewing, and it is the males that do it. With their sharp bills, the males pierce holes in tough evergreen leaves and thread grassy or spider-web threads through them that draw the leaves together. Then they tighten and knot the ends.

For or five large leaves, or a dozen or more smaller ones, gathered in this way become a bag, which one or both partners line with soft grass or vegetable fibers. Some kinds of tailorbirds stitch on a roof to keep the rain out. Then the hens take over, laying and incubating three or four eggs. Asian tailorbirds work in pairs, a male and hen cooperating over each nest. Australian tailorbirds have different rules – each male builds several nests, in which different hens lay eggs and raise their young.

Stick-on camouflage

This caterpillar lives in a bag of its own silk, cutting up leaves and twigs to decorate its home. The larva of a moth, or bagworm, it feeds on leaves, extending the bag as it grows and fattens, and sticking on debris that helps to camouflage it (p. 26). Eventually it forms a pupa or chrysalis, which becomes an adult moth.

Attaching eggs

Lice, tiny parasitic insects, live on the head and body hairs of mammals, feeding by sucking their host's blood. They reproduce by laying eggs, which they stick firmly to the host's hairs. The eggs, called nits, hatch in seven or eight days. Lice on humans thrive among dirt and overcrowding, and carry a disease, typhus, that can be fatal.

Saliva glue

Beautifully streamlined, swifts whiz through the skies with mouths wide open, catching insects in flight. Their nests are a few feathers or straws, held together with a specially produced sticky saliva. This African palm swift has stuck its nest to a vertical palm frond, using an extra squirt of glue to hold the two eggs in place.

Froth-blowers

These tiny patches of white or greenish foam, called "cuckoo spit" or "spittlebug spit," have nothing to do with cuckoos or spitting. They are made by plant-sucking insect larvae, for which spittlebug is a good general name. The larvae blow air through their excreted wastes, making the sticky foam patches in which they hide from predators.

Packagers

Wrapping in silk

Female black widow spiders spin irregular, untidy webs in caves and undergrowth. When an insect becomes entangled, the spider feels the vibrations and runs forward, wrapping

the insect in a network of silk from her spinning glands. Injecting poison to kill it and saliva to digest it, she sucks out the liquid body contents.

Horned eggs

Dogfish package their eggs in curiously shaped horned cases, which they lay among seaweed in shallow water. Each female lays a dozen or more eggs. Each egg has four curling tendrils at the corners, which catch in the weeds and anchor it in place. The parents take no further interest. The baby dogfish emerges, and the case drifts away, its work done.

Mother birds, reptiles, amphibians, fish, spiders, and insects are all "packagers," developing inside themselves the packages called eggs, in which their young will grow. Each egg is produced in a kind of assembly line. An egg contains an embryo, or young animal, usually too small to see, plus the food it will need for growing. Last to be added is the shell or casing that, after the egg is laid, keeps out weather and predators. As the egg develops, the food disappears, its contents reshaped into muscles, skin, blood, and other structures of the growing embryo.

Does this mother hen know she is a packager? No, she has no control over the eggs or what went into them. As they appear, she responds by sitting on them, holding them closely under her for 21 days to keep them safe and warm. Then the hard, chalky shells crack and the chicks are emerge. The hen clucks softly, the chick cheeps in reply. A new stage of motherhood begins.

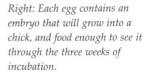

Right: Each egg contains an embryo that will grow into a chick, and food enough to see it through the three weeks of incubation.

Chrysalis

Here is another kind of package, made by the caterpillar of a peacock butterfly. Emerging from a tiny egg, the larva has eaten for three weeks, growing and getting fat. Now it has found a sheltered corner and grown a hard shell, becoming a chrysalis. Within its package it completely changes, or metamorphoses, emerging after a few weeks as a butterfly.

Leaf-cutter

Watch this small gray bee land on a rosebush and carve pieces from the edge of a leaf with its jaws. A leaf-cutter bee collects wrappers for its eggs. Leaf-cutter queens pack their eggs into long narrow tunnels in the soil. Each egg is wrapped individually, with enough nectar and pollen to feed the larva.

Defense Masters

The world of animals includes many predators, ready to pounce and eat the unwary. Every animal has ways of defending itself, either by running, or by facing its attacker with defiance and counter-attack. Big animals have least to fear – size and weight protect them from lesser animals. But what of the smaller ones? Many use interesting tricks to deter or outwit predators.

This African cobra faces a fierce, angry mongoose. The mongoose weaves like a boxer from side to side, seeking to grab the snake's throat. The cobra has spread its hood, making its throat more difficult to grasp. It too is weaving. If it can grab the mongoose, its paralyzing poison will win. If it misses, the mongoose will make the first bite, and possibly cripple it. The snake has one extra trick. It is a spitting cobra, and can throw a fine jet of poison from its mouth, enough to blind and even paralyze the mongoose. This handy natural weapon may well end the fight.

Opposite page: Attacked by a mongoose, the cobra defends itself by moving constantly, altering its shape, threatening to strike – and spitting poison.

Chemical warfare

This bombardier beetle, found in northern America, has small wings, useless for flight, and so has developed a different form of defense. When it is threatened, a number of chemicals are produced inside a special cavity in the beetle. As these chemicals mix, a powerful reaction takes place and a strong-smelling, boiling-hot mixture sprays out of the abdomen, stunning bird or mammal predators.

Raising a stink

Mammals produce scents from glands under their skin. Some, like skunks, store the scent and eject it in self-defense, making a concentrated smell that other animals hate. Distantly related stink badgers have a milder, but still very strong scent. Released when they are alarmed or excited, the smell confuses enemies and increases the badgers' chances of escape.

Jet of oil

This snow petrel is incubating on a ledge in Antarctica. The brown skua, hoping to snatch its egg, approaches, and the petrel clops its bill in warning. If the skua comes closer, the petrel will eject a stream of foul oil from its stomach, which the skua will find very hard to clean off.

Defensive spit

South American vicuñas belong to the camel family. Camels of Asia and Africa are notorious for spitting at things they don't like, and vicuñas are no better. Like cows and sheep, they regurgitate their food, bringing it back from their stomachs to chew a second time. Spitting makes a useful defense weapon.

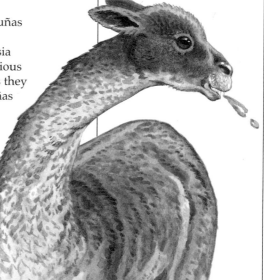

Sailors and Divers

Opposite page: Largest of the toothed whales, sperm whales live in all the world's oceans, feeding on fish and squid that they catch in deep water.

People go to sea in well-built ships. Sea animals have to be built for life in saltwater. Some of the world's largest animals live in the ocean. Among the biggest, this bull sperm whale measures over 60 feet (18 m) long and weighs as much as ten elephants. On land, it would be hopeless – too big and heavy even to breathe, and incapable of moving around. In the water, it floats comfortably at the surface, swims fast when it needs to, and migrates each year from tropical to polar seas and back. It can dive, too. The deep-water fishes and squid that this whale feeds on live far below the surface. Whales are mammals, so they must come up to breathe between dives. Nevertheless, the sperm whale can hunt at depths of 3,000 feet (900 m) or more, and can stay down for an hour or longer.

Walking on water

Here is an insect that walks on water. It is a pond-skater, a beetle that skims over the water's surface on four long and two short legs, looking for scraps of food. Why doesn't it sink? Well, surface tension, the force that holds the water surface together, is strong enough to support its weight.

Built for swimming

Emperor penguins breed on ice floes off the Antarctic coast, feeding in the cold, dark waters beneath. They have dense overlapping feathers and thick blubber to keep out the cold. On the ice, they waddle, ungainly like overweight people. But they are built for life in the water. They swim beautifully and quickly, and dive to depths of over 1,200 feet (370 m).

Hunting under the ice

Arctic bearded seals spend about half their lives in water, the other half on sea ice. Some may never visit land at all. They hunt in near-freezing coastal waters, diving to 300 feet (90 m) or more. In the dark below the ice floes, the bristly beard helps them feel for clams and fish. Thick blubber, or fat, beneath the skin helps keep them warm.

Swimming snakes

Sailors once told tales about sea serpents – fearful monsters longer than ships. Real sea snakes are much smaller, seldom more than 5 feet (1.5 m) long, the largest up to 10 feet (3 m). They live in the tropical Indian and western Pacific oceans. Permanently in water, their bodies are flattened from side to side, enabling them to swim by using their tail as a paddle. They are highly poisonous.

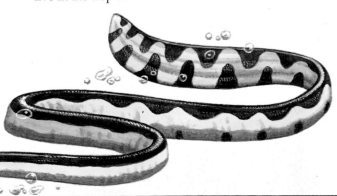

Chompers and Grinders

Left: Sawfish carry a double row of flat, very sharp teeth on either side of their upper jaw, for use in feeding and possibly fighting.

Animals' teeth vary greatly in both shape and in what they do. You can see some of the strange teeth and tusks on these pages. Human beings have four different kinds of teeth – incisors for cutting, canines for ripping and tearing, and premolars and molars for crushing and grinding. Some other mammals and some reptiles have teeth that are far more specialized. Fish teeth tend to be simpler, but there are some that are very unusual. This sawfish, distantly related to sharks, has rows of flat, razor-sharp teeth like a two-edged saw along its upper jaw, and a second set of crushing teeth inside its mouth. The saws, which can be up to 10 feet (3 m) long, rake the seabed for clams, and swish through shoals of fish, crippling and damaging as many as possible. Then the fish sucks the debris into its mouth, grinding, crushing, and swallowing. Sawfish use their teeth for fighting, too.

Strong jaws

People call them "laughing hyenas" – so do they have a strong sense of humor? No, just a call that sounds like a laugh. For their size, hyenas have the strongest jaws of any mammal. With huge jaw muscles and a set of spiking, cutting, and grinding teeth, they can finish off anything they kill, bones and all.

What teeth?

What sort of teeth do you need to eat termites? Ask this aardvark, which eats them all the time. Aardvarks feed by tearing down termite nests, sticking in their blunt noses, and licking up the insects. They don't have any teeth in the front, and just a few simple grinders in the back. The hairs in their nostrils keep termites out.

Beaked whale

A whale with a beak? Well, it looks like a beak, but is actually a long narrow jaw. Baird's beaked whales eat squid, swallowed whole. They get by with no teeth on the upper jaw, just two on either side in the lower jaw. Males especially are covered with long scratches, suggesting that the teeth are for fighting.

Chisel teeth

A rodent native to South America, the coypu is now widespread across Europe and Asia. At first sight, it looks like a nightmare rat, with huge front teeth bloodstained from its last victim. It is, in fact, a gentle, shy animal. The teeth are naturally red, and used for gnawing the roots, reeds, and other vegetation that make up its diet.

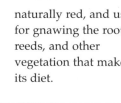
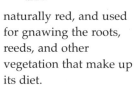

Drillers

Good things don't always come easily. Both herbivores (plant-eating animals) and carnivores (meat-eaters) often have to work for their food, digging, tearing, or drilling through protective covers to get to the goodies inside. This woodpecker (opposite page) feeds on grubs – the larvae of beetles – and other insects that live in the soft, well-rotted wood of old trees. To find them, it often has to drill and chisel through much harder wood, using that strong bill.

As a bill it looks pretty straightforward – a sharp point for drilling – but there's more to it than that. The point has a hard, bony core, and the whole skull is strengthened and fitted with shock-absorbers for constant hammering. The tongue that pulls the grubs out is twice as long as the bill, supported by long, thin bones, and operated by muscles that curve right around the back of the skull.

Drilling for sap

Here is another kind of tree-drilling bird, the yellow-bellied sapsucker. As their name suggests, sapsuckers are after sap – the natural juices of trees – rather than insects. They drill through the bark of young, lively trees to tap the flow beneath. Sap is most abundant in spring. Later in the year, when it flows more slowly, they feed mainly on insects.

This pileated woodpecker is drilling into soft wood with its bill. Its skull is especially cushioned to absorb the constant hammering.

Shellfish driller

A sea snail with a spiny, trumpet-shaped shell, this murex whelk crawls over the seabed in search of prey. It feeds through a trunk-like proboscis (mouth) that sticks out from the shell opening. The proboscis drills through the hard shells of oysters and other mollusks, killing them, and sucking out their contents.

Shells that drill

Long, slender, and slimy, shipworms are mollusks, related to oysters, with two tiny shells on the front end that drill into wood. Usually it is the wood of mangrove trees, growing along tropical shores. But they also burrow into dock piles, driftwood, and the timbers of wooden ships, weakening them and causing immense damage.

Rock-borer

Drilling through timber can be tough, but this piddock, or rock-borer, similar and closely related to the shipworm (top right), drills through rocks. Cased like a mussel or clam in a double shell, it drills by small rocking movements of the shells. Rasping the rock away creates a circular tunnel, in which the piddock spends its life.

Carpenters and Woodworkers

Left: This brown gardener bowerbird of Australia has laid out a lawn and garden, and built a wooden summerhouse – all to attract mates.

Paper nest

This wood pulp nest houses over a hundred German wasps, the offspring of a single queen. Emerging from sleep in late winter, the queen began chewing slivers of wood, molding them into the paper-thin sheets that form the brood chamber. On papery shelves within the chamber she laid her eggs, raising a brood of workers and her own successor.

Cutting bark

Pull away the bark of an old tree, and you may find a pattern like this underneath. It tells the story of a beetle family. A mother bark beetle made the central avenue, eating the wood and laying a batch of eggs as she went. Larvae that hatched from the eggs ate their way along the side avenues, eventually emerging as adults.

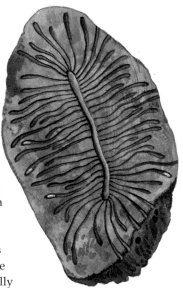

There are 17 different kinds of bowerbirds in Australia and New Guinea, all with strong instincts for building and decorating. A bower is a kind of showcase or platform, created by a male at the start of the breeding season. Different species have different ideas on how to decorate them. Some collect colored stones, shells, scraps of paper, or flowers. Some are woodworkers, circling their plots with neat boundary fences, creating avenues and arcades of twigs. Gardener bowerbirds trim the undergrowth to create patches of lawn several feet across, decorate them with leaves, flowers, and berries, and set up towers of carefully selected twigs. What for? As a stage on which the male can dance, to attract the hens.

This brown gardener bowerbird has laid out his platform, created a lawn from moss, brought in flowers for decoration, and built a 3 foot (1 m) high cottage of cut-to-length timbers. If that won't attract hens, what will?

Working through wood

A sawfly (below right) is a big black-and-yellow flying insect with what looks like a huge stinger on its tail. It looks very dangerous. But, sawflies are the mildest of insects. You can handle one safely, but be sure it's a sawfly, not a hornet. The stinger is a tube that works like a saw and drill, and is used to cut into timber, so the insect can lay eggs deep inside the wood.

Chiseling away

Rats carry a pair of sharp chisels – their two front teeth. Squirrels, chipmunks, and other small rodents use these teeth for gnawing nuts and soft shoots. Rats that live in timber-frame buildings and warehouses chew holes in the wood so they can get around, chiseling their way from room to room, cupboard to cupboard, and box to box in search of food.

Built-in Tool Users

Nutcracker
A nut is an embryo and a source of food, wrapped in a protective shell. When you crack a nut, it's the kernel that you're looking for, rich in oils, proteins, and minerals. The bill of this Clarke's nutcracker is a multipurpose tool for pulling pine nuts out of their cones, cracking the shells, and extracting the kernels.

Why don't birds have hands? Because the bones and muscles of the forelimbs are built into their wings – they cannot have both. Where we use hands, they use their bills. Look at the different kinds of birds on these pages and elsewhere in the book, and you'll see how their bills are modified in different ways for different purposes. There are long, narrow bills like tweezers, short stout ones like pliers, some used for drilling, some bent, some straight, some strong enough to crack nuts. All are examples of good engineering. If they were not, they wouldn't work. This skimmer (opposite) – a water bird closely akin to gulls and terns – has a bill that looks like a serious mistake. The lower mandible (jaw) is half as long again as the upper. Why is that? You'll understand when you watch a skimmer feeding. It skims low over the water with that long, lower mandible trailing, snapping up fish, crabs, and other surface-living foods.

Filter and strainer
Long-legged water birds, flamingos (seen also on page 15) feed on tiny particles of plant life that float in shallow surface waters of tropical lakes. To collect the food, a flamingo turns its head upside-down and, using its tongue as a plunger, pumps the water through a filter in its bill, straining out the plant cells.

Better bent
A long, thin bill with a curve in it – is this damaged? No – just a special tool for a special job. The wrybill of New Zealand finds insects in the damp earth under stones. A slightly bent bill (bent always to the right, by the way) hauls the insects out from under the rocks more easily than a straight one.

Opposite page: A skimmer uses the lower part of its bill to snap up food from the surface of the water.

Sword-billed hummingbird

Sickle-billed hummingbird

Sword and sickle
Hummingbirds feed by hovering in front of flowers and lapping up the nectar with their long, thin tongues. Flowers come in a range of different shapes, so hummingbird bills also need to be different shapes and sizes to be able to reach their food source. These two hummingbirds have straight and curved bills, for dealing with straight and curved flowers.

Tool Makers and Users

People used to say that man is the only tool-using animal. Then naturalists found several other kinds of animals, like the Galapagos Islands finch (opposite page) and the chimpanzee (left), use naturally occurring tools. So then they said man is the only animal that makes tools. But what about a finch that chooses the right kind of spike, or a chimpanzee that finds a stick of the right length? Aren't they making tools with a purpose?

Woodpecker finches of the Galapagos Islands feed on the larvae of beetles that burrow in old wood. The finches' bills are too thick to poke into the tiny holes left by the beetles. So, instead, they find long spines of prickly pear and other cactuses, and use them to probe the larvae out.

The more we look around, the more animals we see using tools, and sometimes making or selecting them for particular purposes.

Working it out
We find close-to-human behavior in monkeys and apes, the animals whose ancestry we think is closest to our own. Chimpanzees, probably closest of all, seem the most ingenious. This captive chimpanzee has piled up three boxes to make a stepladder, and is poking the food with a stick, to reach the bananas above its head.

Woodpecker finches of the Galapagos Islands pick up cactus spikes and use them to extract burrowing insect larvae from timber.

Spikes for storage
Red-breasted shrikes feed on beetles, nestling birds, and other small animals. Some days these are so plentiful that the birds cannot eat fast enough. The next day the food may be scarce. So a shrike continues to catch food while it is there, storing it temporarily on spikes. Thorns are the usual spikes, but a barbed wire fence serves just as well.

Breaking snail shells
Snails live in shells, and a shell is tough and awkward for a thrush's bill to penetrate. So the thrush bangs the shell against the ground. On soft ground it doesn't work, but against a stone the shell breaks every time. If you have thrushes in your garden, you may find their "anvils" – stones they use every day for cracking snail shells.

Stone hammer
This Pacific sea otter has found a sea urchin and a stone from the seabed. Too spiky to bite, the urchin has to be cracked. So the otter lies on its back in the water, with the urchin on its chest, hammering it on the stone. When it breaks, the otter will scoop out and eat the soft contents.

Temperature Controllers

Keeping warm

Emperor penguins live among Antarctic snows and swim in icy seas. Thick feathers, waterproof and windproof, keep them warm, with a downy undershirt and an inch-thick layer of fat beneath the skin. In

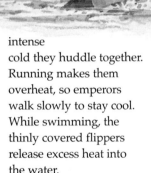

intense cold they huddle together. Running makes them overheat, so emperors walk slowly to stay cool. While swimming, the thinly covered flippers release excess heat into the water.

Sharing warmth

Musk-oxen spend their lives on Arctic tundra, exposed to the coldest winds. A dense mantle of woolly fur up to 1 foot (30 cm) thick covers the body, with an outer covering of longer guard hairs that shed snow and keep the oxen dry. Always living in groups, they huddle together for warmth. A lone musk ox would probably not survive the harsh Arctic winter.

Birds and mammals, including humans, are warm-blooded, with body temperatures close to 98-99°F (about 37°C). Are angry people hot-blooded and cool people cold-blooded? No, we all have built-in controls that keep our body temperature very constant. If it rises or falls even one or two degrees, we feel unwell, feverish, or worse. Chemical changes in our muscles and other organs keep us warm. If we run, or sit in hot sunshine, our body temperature starts to rise, and we have to cool down. If we stand around in the cold, our temperature starts to fall, and we have to warm up.

When humans overheat, our sweat glands produce water that evaporates and cools us. Furry animals have fewer sweat glands, so they need other ways of cooling. In hot weather this eland, a large African antelope, rests in the shade and stays still. If it overheats, those long horns become radiators. Under the horny surface runs a network of blood vessels that flush with blood. This cools the blood, and

Licking

It is hard for Australian wallabies (right) to keep cool in summer. Grazing mainly in the morning and evening, they lie in the shade from noon onward, often licking their paws and tail.

Keeping clean? No, cooling off. These areas are less furry than the rest of their body. Saliva evaporates quickest from them, helping to cool them down.

Panting

How does a hot dog cool down? It pants (above). With mouth open and well moistened by saliva, it breathes at eight to ten times the normal rate. The saliva evaporates, cooling the tongue and the blood flowing through it, and so cooling the dog. Few other animals pant like this. If you try it, you'll soon become dizzy.

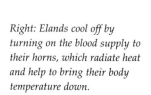

Right: Elands cool off by turning on the blood supply to their horns, which radiate heat and help to bring their body temperature down.

Index

A
aardvark 37
African antelope 46-7
albatrosses 21
amphibians 16
anchors 22
ant lion 9
antelope 46-47
anvils 45
archerfish 18-19
armadillo 25
armor plating 24-5
Australian desert termites 16

B
badgers 10, 33
Baird's beaked whale 37
bark beetles 41
bark cutting 41
beaked whale 37
beaks/bills 39, 42-3
bearded seals 34
bears 10-11
beavers 6-7, 16
bee, leaf-cutter 30
bees 30
beetles 9, 41
birds 9, 10, 12-13, 14-15, 21, 26-7, 28-9, 33, 34, 38-9, 44-5
black-headed gulls 9
black widow spiders 30
blubber 34
boa 25
bowerbirds 40-1
bowers 40-1
brown skua 33
builders 16-17
bumble-bees 10
burrows/burrowers 10-11
butterflies 19, 30

C
cactus spikes 44-5
caddis flies 22
camels 33
camouflage 19, 26-7, 29
canines 37
carpenters 40-1
caterpillars 30
cave 10-11
chaffinch 12
chemical warfare 31
chimpanzees 45
chipmunk 8-9
chisel teeth 37
chitin 25
chompers 36-7
chrysalis 30
Clarke's nutcracker 42
cobra, African 32-3
cobra, Indian 10
cobwebs 12-13
conches 25
coral reefs 16
cormorants 14-15
coypu 37
crustaceans 24-5
cuckoo-spit 29

D
dam-builders 6-7, 16
defense 32-3
diggers 8-9

dogfish eggs 30
drillers 38-9

E
earthworms 9
eggs 10, 12, 15, 26, 29, 30-1, 41
eider ducks 26
eland 46-7
emperor penguins 34, 46
Eskimo 10

F
feathers 26, 46
finches 44-5
fish 18-19, 21, 25, 36-7
fishing nets 22
flamingos 15, 42
flying 20-1, 22
flying fish 21
flying squirrels 19
food 9, 18-19, 22, 26, 29, 30, 33, 34-5, 36-7, 38-9, 42-3, 44-5
frog, blacksmith 16
frog, flying 20-1
froth-blowers 29
fur 46

G
Galapagos Islands finches 44-5
gardener bowerbird 40-1
gliding 20-1
grinders 36-7
grubs 39
guard hairs 46
gulls 9

H
hammers 45
harvest mouse 12
hens and chicks 30-1
holts 10
hornbills 15
horns 46-7
humans 6, 10, 37, 46
hummingbirds 12-13, 42
hydraulics 19

I
incisors 37
Indian cobra 10
ink clouds 19, 26
insects 9, 10, 21, 26, 29, 30, 34, 41
Inuit 10

J
jaws, strong 37
jet engineers 18-19

L
larvae 9, 22, 29, 41, 44-5
laughing hyenas 37
leaf-cutter bees 30
lice 29
licking 46
lobsters 24-5
lodges 6-7, 16

M
malimbes 12
mammals 10-11, 12, 25, 33, 34, 37, 45,46
mason bee 26
mice 12

molars 37
mollusks 19, 22, 39
monarch butterflies 19
mongoose 32-3
monkeys and apes 45
moths 9, 29
murex whelk 39
musk oxen 46
mussels 22

N
nests 10, 12-13, 14-15, 16-17, 26, 28-9, 41
nits 29
nutcracker (bird) 42
nuthatches 15
nuts 8-9, 42

O
octopi 19, 26
oil jet 33
oven birds 16-17
oviduct 30
oysters 19

P
packagers 30-1
palm swift 29
panting 46
parachutes 21
peacock butterfly 30
penguins 34
piddock 39
pileated woodpecker 38-9
pit traps 9
plasterers 14-15
polar bears 10-11
pond-skater 34
premolars 37
prey, catching 9
propulsion 19
protective plates 24-5

Q
queen bee 30
queen conch 25

R
rats 41
rays 25
red-breasted shrikes 45
red-headed malimbes 12
rock-borer 39
rodents 6-7, 8-9, 12, 19, 37, 41
rooks 9

S
saliva 26, 29, 46 see also spit
saliva glue 29
sand martins 10
sap, drilling for 39
sapsuckers 39
sawfish 36-7
sawfly 41
scales 25
scallops 19
scent, defensive 33
sea otter 45
sea serpents 34
sea urchins 26, 45
seals 34
seaweed 14-15, 26
setts 10

sharks 25
shellfish see mollusks
shells 19, 25
shipworms 39
shrikes 45
sickle-billed hummingbird 42
silk spinning 22
silkworms 22
skeleton, exterior 24-5
skimmer 42-3
skin shedding 25
skunks 33
smoke screens 19, 26
snails 25, 39, 45
snakes 10, 25, 34
snow cave 10-11
snow petrel 33
sperm whales 34-5
spiders 6, 22-3, 29, 30
spider's web 6, 12, 22-3
spinners 22-3
spit 29, 33
spitting cobra 32-3
spittle-bug 29
sponges 19
stink badgers 33
surface tension 34
swallows 15
swifts 29
swimming pools 16
sword-billed hummingbird 42

T
tadpoles 16
tailorbirds 28-9
teeth 36-7, 41
temperature control 46-7
termites/termite mounds 10, 16, 37
thrushes 45
tool makers and users 44-5
tool users 42-3, 44-5
tower-builders 16
tree frog 20-1
tunnels 9

U
underground living 9

V
vicuñas 33
village weaverbirds 12

W
wallabies 46
wandering albatrosses 21
warrens 10
wasps 41
weaverbirds 12
weavers 12-13
webs 22-3, 30
whales 34-5, 37
whelks 39
white-eared hummingbird 12-13
woodpecker finches 44-5
woodpeckers 15, 38-9
woodworkers 40-1
worms 9, 22
wrybill 42
wryneck 26-7

Y
yellow-bellied sapsucker 39